Gwen M. Nuelands
3456 W. 16th Ave
Vancouver 8 B.C.

AT HAMPTON COURT

THE THAMES

DESCRIBED BY G. E. MITTON
PICTURED BY E. W. HASLEHUST

BLACKIE & SON LIMITED
LONDON AND GLASGOW

BLACKIE & SON LIMITED
66 Chandos Place, London
17 Stanhope Street, Glasgow

BLACKIE & SON (INDIA) LIMITED
Warwick House, Fort Street, Bombay

BLACKIE & SON (CANADA) LIMITED
Toronto

BEAUTIFUL ENGLAND

Chester.
Windsor Castle.
Rambles in Greater London.
In London's By-ways.
The Thames.
The Peak District.
The Cornish Riviera.
Oxford.
Canterbury.
Shakespeare-land.
Exeter.
Dickens-land.
Through London's Highways.
The Heart of London.
Bath and Wells.
Winchester.
Dartmoor.
Cambridge.
York.
The English Lakes.

BEAUTIFUL SCOTLAND

Loch Lomond, Loch Katrine and the Trossachs.
The Scott Country.
Edinburgh.
The Shores of Fife.

BEAUTIFUL IRELAND

Ulster.
Munster.

Printed in Great Britain by Blackie & Son, Ltd., Glasgow

LIST OF ILLUSTRATIONS

WINDSOR

When the American wondered what all the fuss was about, and "guessed" that any one of his home rivers could swallow the Thames and never know it, the Englishman replied, he "guessed" it depended at which end the process began; if at the mouth, the American river would probably get no farther than the "greatest city the world has ever known" before succumbing to indigestion!

With rivers as with men, size is not an element in greatness, and for no other reason than that it carries London on its banks the Thames would be the most famous river in the world. It has other claims too, claims which are here set forth with pen and pencil; for at present we are not dealing with London at all, but with that river of pleasure of which Spenser wrote:—

Along the shores of silver-streaming Themmes;
Whose rutty bank, the which his river hemmes,
Was paynted all with variable flowers,
And all the meades adorned with dainty gemmes,
Fit to deck mayden bowres and crowne their paramoures,
Against the brydale day which is not long,
Sweet Thames! runne softly till I end my song.

Oddly enough, this is one of the comparatively few allusions to the Thames in literature, and there is no single striking ode in its honour. It is perhaps too much to expect the present Poet Laureate to fill the gap, but certainly the poet of the Thames has yet to arise.

Besides Spenser, Drayton makes allusion to the Thames in his *Polyolbion*, using as an allegory the wedding of Thame and Isis, from which union is born the Thames; and in this he is correct, for where Thame and Isis unite at Dorchester there begins the Thames, and all that is usually counted Thames, up to Oxford and beyond, is, as Oxford men correctly say, the Isis. Yet by custom now the river which flows past Oxford is treated as the Thames, and when we speak of our national river we count its source as being in the Cotswold Hills.

Other poets who refer to the Thames are Denham, Cowley, Milton, and Pope. In modern times Matthew Arnold's tender descriptions of the river about and below Oxford have been many times quoted. Gray wrote an *Ode on a Distant Prospect of Eton College*, in which he refers to the "hoary Thames", but the lines apostrophizing the "little victims" at play are more often quoted than those regarding the river.

The influence of the Thames on the countless sons of England who have passed through Eton and Oxford must be incalculable. It is impossible to mention Eton without thinking of Windsor, the one royal castle which really impresses foreigners in England. Buckingham Palace is a palace in name only, its ugly, stiff, stuccoed walls might belong to a gigantic box, but Windsor, with its massive towers and its splendid situation, is castle and palace both. Well may the German Emperor envy it! It carries in it something of the character of that other William, the first of the Norman Kings of England, who saw the possibilities of the situation, though little of the castle as we see it is due to him. The mass of it is of the time of Edward III, and much of it was altered in that worst era of taste, the reign of George IV. Windsor has come scatheless out of the ordeal; the fine masses of masonry already existing have carried off the alterations in their own grandeur, and the result is harmonious.

Many and many a tale might be quoted of Windsor, but these are amply told in *Windsor*

Castle by Edward Thomas, the volume which follows this in the same series. Here we must be content with quoting only four lines from *The Kingis Quhair*, the great poem of King James I of Scotland, who spent part of his long captivity at Windsor. By reason of this poem James I ranks as high among poets as among kings; in it he speaks of the Thames as—

> A river pleasant to behold,
> Embroidered all with fresh flowers gay,
> Where, through the gravel, bright as any gold,
> The crystal water ran so clear and cold.

Windsor is the only royal palace, still used as such, which remains out of the seven once standing on the banks of the Thames. Few people indeed would be able to recite offhand the names of the others. They are all below Windsor. The nearest to it is Hampton Court, chiefly associated with William III, though it was originally founded by the tactless Wolsey, who dared so to adorn it that it attracted the unenviable notice of Henry VIII. Little was it to be wondered at, since the Court was described by Skelton as—

> With turrettes and with toures,
> With halls and with boures,
> Stretching to the starres,
> With glass windows and barres;
> Hanginge about their walles,

Clothes of gold and palles
Fresh as floures in May.

Skelton also wrote a satire beginning:—

Why come ye not to court?
 To whyche court?
To the Kynge's Court
Or Hampton Court?
The Kynge's Court
Should have the excellence,
But Hampton Court
Hath the pre-eminence
And Yorkes Place,

which was like pouring vitriol into the mind of such a man as Henry. When Wolsey entertained the French ambassadors at Hampton, "every chamber had a bason and a ewer of silver, some gilt and some parcel gilt, and some two great pots of silver, in like manner, and one pot at the least with wine or beer, a bowl or goblet, and a silver pot to drink beer in; a silver candlestick or two, with both white lights and yellow lights of three sizes of wax; and a staff torch; a fine manchet, and a cheat loaf of bread". No wonder the King's cupidity was aroused. It was not long before the great Cardinal was forced to make a "voluntary" gift of his beloved toy, as he had also to do with another noble mansion which he "made" by Thames side—Whitehall, formerly known as York Place, because held by the Archbishops of York. When Wolsey was told the King required this, he said with truth: "I know that the King of his own nature is of a royal stomach!"

On leaving Hampton the great prelate was allowed to go to the palace at Richmond. One wonders if he rode from Hampton to Richmond, only a mile or two by the river bank, on that "mule trapped altogether in crimson velvet and gilt stirrups". Of the thousands who use that popular towpath does one ever give a thought to the Cardinal thus setting his first step on his tremendous downward descent?

It was while he was at Hampton that the news was brought to Henry of the death of his old favourite at Leicester Abbey. Henry, standing in a "nightgown of russet velvet furred with sables", heard the news callously, and only demanded an account of some money paid to the cardinal before his death; not a qualm disturbed his self-satisfaction. Such is the most picturesque reminiscence of Hampton, and others must stand aside with a mere reference; such events as the birth of Edward VI, which occurred here; the "honeymoon" of bitter, loveless Mary and her Spanish husband; the imprisonment of Charles I for three months. Melancholy ghosts these; but they do not haunt the main part of the palace, for that was built later by Wren, acting under orders from William III, to imitate Versailles.

This incongruity of style must have sorely puzzled the much-tried architect, who has, however, succeeded admirably in his bizarre task.

But of all the picturesque and romantic associations with palaces, those connected with Richmond are the most interesting. Only a fragment of the building now remains. After many vicissitudes, including destruction by fire at the hands of Richard II —who, like a child rending a toy which has hurt him, had it destroyed because the death of his wife occurred here—it was rebuilt by Henry VII, the first to call it Richmond, whereas before it had been Sheen. It is much associated with the eccentric and forceful Tudors, who, whatever their faults, had plenty of ability, and of that most valuable of all nature's gifts, originality. It is said that in a room over the gateway took place the death of the miserable Countess of Nottingham, who confessed at last that she had failed to give to Elizabeth the ring which the Earl of Essex had sent to her in his extremity; whereupon the miserable queen exclaimed: "May God forgive you, for I never can". The unhappy Katherine of Aragon, and still more unhappy Queen Mary, spent bitter days at Richmond.

How different is Kew, a palace in name only, a snug red-brick villa in appearance, where the most homely of the Hanoverian kings played at being a private gentleman! The other royal palaces—Westminster, Whitehall and the Tower—belong to the London zone, a thing apart, just as London is now itself a county, an entity, and not merely a city overflowing into neighbouring counties.

Not only for its palaces is the Thames famous, the monks made excuse that Friday's fish necessitated the vicinity of a river, but in reality they knew better than their neighbours how to choose the most desirable localities. Note any exceptionally beautiful situation, any celebrated house, and ten times to one you will find its origin in a monastery. The monasteries which dotted the shores of Thames were frequent and lordly. To mention a few of the most important, we have Reading, Dorchester, Chertsey, Abingdon, and an incomparable relic remaining in the magnificent abbey church at Dorchester, with its "Jesse" window, which draws strangers from all parts to see the tree of David arising from Jesse and culminating in the Christ.

Nowadays many besides monks have discovered the desirability of a river residence; too many, in fact, for a house with the lawn of that unrivalled turf, smooth as velvet, bright as emerald, which grows only by Thames side, commands a rent out of reach of all but the well-to-do. How beautiful such river lawns may be can be judged only at the

RICHMOND

time when the crimson rambler is in its glory, flinging its rose-red masses over rustic supports, and finding an extraordinary counterblast of colour in the striking vermilion of the geraniums which line the roofs of the prettily painted houseboats anchored near. A houseboat is not exactly a marvel either of comfort or cheapness, but as a joyous experience it is worth the money. You see them lying up in lines by Molesey and Richmond out of the season, dead lifeless things, with weather-stained paint and tightly shut casements. How different are they in the summer, resplendent in blue and white, lined by flowers and vivified by men in flannels and girls in muslin frocks, with parasols like flowers themselves; then the very houseboat seems alive.

Of all the notable houses which are passed in following "the silver-winding" way of the Thames two cannot be overlooked, because, being perched in lordly situations, they command great vistas of the river. The first is Cliveden, standing high above the woods and facing down the river to Maidenhead. The present house dates only from the middle of the nineteenth century. It has had two predecessors, both destroyed by fire. The first one was built by "Steenie", first Duke of Buckingham, Charles I's favourite. His gay, arrogant life, which came to a fitting end by the assassin's knife, was carried on at

Cliveden with unbridled licence and extravagance. His wardrobe for the journey to Spain with Charles, when Prince of Wales, consisted of "twenty-seven rich suits, embroidered and laced with silk and silver plushes, besides one rich satten incut velvet suit, set all over, both suit and cloak, of diamonds, the value whereof is thought to be about one thousand pounds". It was to Cliveden the duke brought the Countess of Shrewsbury after he had killed her husband by mortally wounding him in a duel, while she stood by disguised as a page and held his horse.

There is nothing more curious than to discover how young were the principal actors in the dramas of history. After a life full of action, of intrigue, of excitement, the first Duke of Buckingham's career was ended at the early age of thirty-six. He left a son and daughter, and another son, Francis, was born shortly after. This boy is described as having been singularly lovable and handsome. He fought gallantly for his King in the civil wars, and was killed when only nineteen at Kingston-on-Thames, thereby, giving us another riverside association. He stood with his back against an oak tree, scorning to ask quarter from his enemies, and fell covered with wounds.

It was an age of masques and dramas, and Buckingham was the patron of many a poet. Ben

Jonson's masques, performed in costumes designed by Inigo Jones, were popular both with him and the King. In later days Cliveden was the scene of another masque, *Alfred*, written by James Thomson, who was staying in the house as a guest of Frederick, Prince of Wales, then the lessee. This masque itself is long forgotten, but it contained "Rule, Britannia!" the national song which thus first made the walls of Cliveden echo, before it echoed round the Empire. The masque was performed at a fête given in the garden, Aug. 1 and 2, 1740. Thomson's connection with the Thames does not end here. It was at the Mall, Hammersmith, that he had previously written *The Seasons*.

Enough has been said of Cliveden to show that not only in situation but in interesting association it takes high rank among river mansions. The other pronouncedly notable high-standing river mansion is Danesfield, above Hurley, built of chalk, and reared upon the great chalk cliffs that here line the river's flood. On the slopes near, in crocus time, the hills shine purple and gold with blossom, resembling a royal carpet spread by someone's lavish hand. The place derives its name from having been the site of a Danish encampment.

But Cliveden and Danesfield do not exhaust the list of fine riverside mansions, though, as they stand so high, they are more conspicuous than most. One of the most delightful and desirable of all the old houses is Bisham Abbey, not far from Marlow, picturesque in itself and redolent of old associations. There is the Bisham ghost, which spreads itself across the river in a thin, white mist which means death to those who try to penetrate it. But the most touching and pitiful tale is of a certain Lady Hoby, one of the family who held the mansion from the time of Edward VI to 1780. She is represented as wandering about in a never-ending purgatory, wringing her hands and trying to cleanse them from indelible inkstains. The story goes that she was condemned thus for her cruelty to her little son, whom, perhaps in mistaken severity, she beat so much for failure to write in his copybooks without blots that the poor child died. It was an age of sternness toward children. We know how Lady Jane Grey suffered, and thought herself "in hell" while with her parents. There were no Froebel schools or Kindergartens then; and it may be the wretched mother was trying to do her duty as she knew it. A curious confirmation of the story was found in the discovery of a number of copybooks behind a shutter during some repairs. The books were of the Tudor period and were deluged in every line with blots!

MARLOW LOCK

Several of the Hobys are buried in the pretty little church, near to which the river laps the very edge of the churchyard. One monument is to two brothers, Sir Philip and Sir Thomas Hoby, and the epitaph on the latter, put up by his sorrowing widow, concludes with the lines:—

> Give me, oh God, a husband like unto Thomas,
> Or else restore me to my husband Thomas.

Like many another disconsolate widow she married again in a few years, so she had presumably found someone who could rank with Thomas! Leland in his *Itinerary* mentions the Abbey as "a very pleasant delightsome place as most in England", and, indeed, so it is, with its grey stone walls, mullioned windows, and high tower rising amid the trees.

Bisham at one time belonged to the Knights Templars, and in 1388 the Earl of Salisbury established here a monastery for Augustinian monks. It was twice surrendered at the dissolution, and the prior, William Barlow, had five daughters, who all married bishops! It seems that the worthy cleric had readily taken advantage of the change which abolished celibacy for the clergy!

Poor Anne of Cleves lived here in retirement, whilst her stepson was on the throne, but she perhaps found the place too quiet after the fierce excitement of being wife to such a monarch as Henry, because it was she who exchanged it with the Hoby family, and went elsewhere. Edward VI seems to have had a liking for sending his relatives here, for he next committed his sister Elizabeth to the care of Sir Thomas, who seems to have treated her well, though she was in fact a prisoner. That she appreciated the beauty of the river scenery is shown by her revisiting the place when she was queen. The great square hall is said with much probability to have been the abbey church, and if so three Earls of Salisbury, the "King-maker" Warwick, and the unhappy Edward Plantagenet, son of the Duke of Clarence, lie beneath the stones. We have lingered a little about Bisham, but few places are so well worth it.

Temple Lock, near by, recalls the Templars, and just above it is another grand old house, Lady Place, also on the site of an abbey. Sir Richard Lovelace, created Baron by Charles I, built here a magnificent mansion, described by Macaulay in his usual rolling style, in his *History of England*. The house, therefore, is younger than Bisham, but the abbey was older, having been founded as far back as 1086. A part of the crypt remains. Here in the dim depths was signed that document which changed the whole course of English history, the invitation to William of Orange to come over and take the throne. The chief conspirator was the second Baron Lovelace, who thus repaid the Stuarts who had ennobled his father!

At Greenlands also, about three miles above Lady Place and Hurley as the crow flies, but more by the winding river, we get another echo of the Civil Wars. We are told that "for a little fort it was made very strong for the King". It belonged at that time to Sir Cope D'Oyley, a stanch Royalist, and when he died his eldest son followed in his steps, and held out even when the Parliamentarians planted their cannon in the meadows opposite and fired across the river. The marks of their balls are said to be still visible on the old walls. Greenlands now belongs to the Hon. W. F. D. Smith, heir to his mother, Viscountess Hambleden. An altogether peculiar case in the peerage this! When the Right Hon. W. H. Smith, First Lord of the Treasury, died, in October, 1891, he just missed the peerage destined for him. A month later it was conferred upon his widow with remainder to her son.

So much for a few of the interesting and romantic associations of the river. But it is not thus the holiday crowds regard it. They seek no meaning in place-names, no historical associations in the grand old mansions passed; to them the river is a playground merely, where every yard of a particular backwater is known, where a favourite boatman reserves a special boat or punt, and where crowds of fellow creatures may be sought or shunned as individual fancy prompts. We might paraphrase Wordsworth and say:

> A place-name on the river's brim,
> A simple name it was to him,
> And it was nothing more.

One might wander from subject to subject while treating of the Thames, finding in each matter enough for a book, indeed the variety of the subjects rivals in scope that famous conversation which ranged "from sealing-wax to Kings". Romance, history, boating, flowers, regattas, and fish are but a few out of the vast number lying ready for choice, and space is limited.

The Thames swans are a feature to be by no means overlooked. They belong to the Crown, the Vintners' and Dyers' Companies, and so ancient are the rights of the companies in this matter that their origin is lost in the mist of antiquity. The annual stock-taking and marking of the swans gives occasion for a pleasant holiday every year about the middle of July; but though the privileged members of the companies and their friends are no longer

MAIDENHEAD BRIDGE

conveyed in "gaily decorated barges", they no doubt enjoy their excursion by steam launch just as much. "Swan-hopping", as it is usually called, is really a corruption of "swan-upping", meaning the process of taking up the swans to mark them according to their ownership. The Vintners used to mark their swans with a large V across the mandible, but this custom, having been protested against in the new spirit of tenderness which has swept over the country, they now give two nicks only, one on each side. The well-known tavern sign "The Swan with Two Necks" is really a corruption of this much-used mark of identification, and should be "The Swan with Two Nicks".

The King is by far the largest owner, and as he has discontinued the custom of having a number of swans and cygnets taken for the royal table, it is probable that swans will increase on the river very rapidly. The swan has always been a royal bird, and in the time of Edward IV no one was permitted to keep swans unless he had a freehold of at least five marks annually. The order for the regulation of the Thames swans, in which this clause appears, runs to thirty clauses, and is a very quaint document. One sentence is as follows: "It is ordained that every owner that hath any swans shall pay every year . . . fourpence to the Master of the Game for his fee, and his dinner and supper free on the Upping Days".

These regulations show that the institution of swans on the Thames is a very ancient one, and the graceful, bad-tempered birds themselves add much to the beauty of the river.

> The swan with arched neck
> Between her white wings mantling, proudly rows
> Her state with oary feet.
>
> —*Milton.*

To light upon another subject. There is in the boating alone enough to occupy many volumes. We might start from the solid punt, furnished with chairs, and shoved out into midstream by three sober snuff-coloured gentlemen; there anchored by its own poles, while the three sit on their chairs in midstream, regardless of the obstruction they form to quicker nimbler mortals, fishing, or rather holding rods, as immovable as themselves, the livelong day. The punt plays such a small part in the whole proceeding, it might well fall outside the boating classification altogether—a mud island would do as well. It has not even the dignity of a ferry boat. From here, through all varieties of broad-beamed, blunt-nosed family boats, to the long slender racing skiffs or the canoe light as a dragon-fly on the wing, we could run the gamut in the Book of the Boat.

The distance between Hammersmith Bridge and Folly Bridge, Oxford, is 103 miles, and the extent and variety of boating on this stretch, to go no lower, is unequalled on any other river in England. The first weir is to be found below Richmond, and the first lock at Teddington. In 1578 there were 23 locks, 16 mills, 16 floodgates, and 7 weirs on the river between Maidenhead and Oxford. Thirty more locks and weirs were added in the next six years. When we find that "the locks were machines of wood placed across the river, and so contrived to hold the water as long as convenient, that is, till the water rises to such a height as to allow of depth enough for the barge to pass over the shallows", we are not surprised to learn that exception was taken to the building of more locks, because so many people had been drowned! The barges were not charged for going up, but only for coming down, which seems a little unreasonable when we realize that "the going up of the locks was so steep that every year cables had been broken that cost £400".

It is curious how easily the river may be divided into "zones", each with its usual habitués quite distinct from those of other zones. Taking it generally, it may be said that the farther from London the more exclusive is the crowd, and this is perhaps because a very large number of Thames lovers live in

of striking interest. Placid green meadows, feathery willows, peaceful cows, and sunny little unpretentious houses are the chief components of almost every view. Weybridge is perhaps the prettiest place, because of the many turnings and windings of the river near it, but Penton Hook, Laleham, Shepperton, and Walton can all claim a quiet prettiness of their own.

Windsor stands by itself, and the influence of Eton is paramount. Then from Bray right on to Marlow we get what must be by far the most popular bit of the whole river.

Bray itself is particularly pleasant, and is associated for all time with the worthy vicar, who was content to turn his coat at the bidding of the party in power sooner than lose his beloved parish. The original vicar lived in the reigns of Henry VIII and his immediate successor, and his mental somersaults were from the Catholic to Reformed Church, and back once more; but the ballad makes him live in the days of Charles II, James II, William, Anne, and George I, a period of over fifty years. As it is rather difficult to get hold of, we may quote part of it here. It runs through all the variations from—

In good King Charles's golden days,
When loyalty no harm meant,

A zealous High Churchman was I,
And so I got preferment.
To teach my flock I never missed,
Kings were by God appointed,
And damn'd are those that do resist
Or touch the Lord's anointed.

When royal James obtained the crown
And Popery came in fashion,
The Penal laws I hooted down
And read the Declaration.
The Church of Rome I found would fit
Full well my constitution,
And had become a Jesuit
But for the Revolution.

. . . .

When George in pudding-time came o'er,
And moderate men looked big, sir,
I turned a cat-in-a-pan once more
And so became a whig, sir.
And thus preferment I secured
From our new faith's defender,
And almost every day abjured
The Pope and the Pretender.

. . . .

For this is law I will maintain
Until my dying day, sir.
Whatever king in England reign
I'll still be Vicar of Bray, sir.

Maidenhead bridges, rail and road, span the river above Bray. Maidenhead is easily accessible by the Great Western Railway main line, and, with Taplow, which comes down to the river on the opposite bank, counts its devotees in thousands. Taplow village is a little distance away, but Skindle's Hotel on that side counts largely in itself as representing Taplow. Not even the sacred Ganges itself could show a crowd more ardent or more gaily clad than this stretch of the river on a fine summer day. The rich ochres and purples of the East are outshone by the soft brilliancy of blues and pinks, the rose-reds and yellows of the gayer sex both in their garments and sunshades. And if the great day, the Sunday after Ascot, be in any way tolerable, Boulter's Lock, all the more sought apparently because of its congestion, is a sight indeed. People come in crowds to stand on the banks and view it as a show.

But all the year round, even in winter, a few visitors may be found in the reach above Boulter's, under the magnificent amphitheatre-like sweeps of the Cliveden woods. The cliff itself rises to a height of 140 feet and is clothed to the very summit. Oak, beech, ash, and chestnut show up against clumps of dark evergreen. The bosky masses are broken here and there by a Lombardy poplar pointing upward, and the whole is wreathed and swathed in shawls of the wild clematis, the woodbine of the older poets, otherwise traveller's joy. Beyond the Cliveden reach is Cookham, beloved of many, with its pretty little church tower peeping over the trees, and opposite is Bourne End, near which is a wide, open reach used as a course for sailing boats. The only

E·W·HASLEHUST

COOKHAM CHURCH

woods that can rival those of Cliveden are the Quarry Woods, opposite Great Marlow, and they lose in effect from not coming right down to the water but sweeping away inland. The Quarry Woods are largely beech and evergreen, and in the autumn the stems, owing to the damp atmosphere, are covered with a vivid green lichen, the thick leaves, turning the burnt red colour peculiar to beeches, not only shine overhead, but make a rich carpet for the ground. Then the woods might well be the enchanted woods of a child's fairy tale, so glorious is their aspect. Between Marlow and Henley, as we have seen, most of the ancient historical associations cluster; within that short space are Bisham, Lady Place, Medmenham, and Greenlands, and the reach of the river is quite pretty enough to tempt people without the added glamour.

Medmenham Abbey is now a carefully composed ruin, with a most attractive-looking cloister close to the river. So well has art aped reality, that it is regarded with much more reverence than many genuinely old buildings which make less display. It is at present a private house, but began its career in the orthodox way as an abbey, being founded about 1200 for Cistercian monks. Few of the thirteenth-century stones can now remain, unless it be as foundations.

A weird and ghostly flavour was imparted to the place by its being chosen as headquarters by the roistering crew of the eighteenth century who called themselves "The Hell-Fire Club", and professed to worship Satan. The leader of the revellers was Sir Francis Dashwood, who succeeded his uncle in the title of Baron le Despencer in 1763. The club motto was *Fay ce que voudras*, and each member tried to outdo the rest in eccentricity. Though they gloried in their wild doings and set afloat many tales which made quieter folk catch their breath in horror, it is probable that, apart from open blasphemy, their proceedings were more foolish than horrible. Once, as a joke, someone sent an ape down the chimney while they were gathered together, and the frightened gibbering creature, soot-begrimed, was mistaken by the terror-stricken revellers for Satan himself.

Not far off is the old Abbey Hotel, beloved of artists, and farther on up the green lane is a curious old house which once belonged to Sir John Borlase, friend of King Charles II, who was visited here by His Majesty on horseback, often accompanied, so tradition goes, by Nelly Gwynne.

Henley, of course, boasts the regatta of the Thames; other regattas there are in plenty, but none can compare with Henley in importance. Its heats are telegraphed abroad, and as a sporting event it ranks only second to the boat race. The regatta is held the first week in July. The course is lined by booms, within the shelter of which every variety of craft is seen wedged together so tightly as to make upsetting a sheer impossibility. Punts worked with canoe paddles are perhaps the most popular, but skiffs and frail Canadian canoes, as well as the solid hired craft of the boat builders may be seen. Gondolas regularly make their appearance, and seem to vanish in between from year to year. It used to be fashionable to wear simple muslins and straws at Henley, but year by year fashion has screwed up things to a higher pitch, until nowadays gowns which, in their elaborate affectation of simplicity, would not disgrace Ascot itself, are to be seen everywhere, especially on the lawns of the clubs which run down to the water behind the waiting craft. The scene is a gay one, and for days before every available room is taken, every available boat hired. The Red Lion—and Henley would hardly be Henley without the Red Lion—could be filled several times over. It was of this inn Shenstone wrote:—

> Whoe'er has travelled life's dull round,
> Whate'er his stages may have been,
> May sigh to think he still has found
> The warmest welcome at an inn.

The whole poem, of which this is a verse, was written on a window of the inn, and though the window was broken the relic is preserved. Charles I stayed at the Red Lion in 1632, on his way from London to Oxford, and a large fresco painting of the Royal Arms, done in commemoration of this visit, was discovered over a fireplace during alterations. Doubtless it had been purposely hidden in the days when Henley was hotly Parliamentarian and striving vainly to subdue poor little Greenlands.

Owing to its position as a sort of halfway house between London and Oxford, Henley enjoys a good deal of society. The great Duke of Marlborough actually furnished a room at the inn that he might frequently occupy it. It is at Henley that the daily steamer stops when running between Kingston and Oxford in the summer months.

Between Henley and Sonning lies the most intricate part of the river bed, and here are the most bewitching reaches. The numerous islets, the backwaters and sheltered nooks, make it a favourite part with boating men.

Wargrave backwater, indeed, is the most famous on the river, and is in summer simply a fairyland of greenery. The entrance, behind a willow-covered island, conveys something of mystery, and as one floats gently along a waterway so narrow that one

HENLEY

could almost touch the banks on either side, with the sun showering down between the meshes of the delicate veil of leaves, one might be sailing into the palace where lies the sleeping princess. Fiddler's Bridge is so low that it is necessary to lie down full length in the boat in passing under it, and two boats meeting must certainly make some arrangement for mutual safety, even if it be not exactly that of the goats in the fable.

Wargrave itself might be taken as a typical Thames-side village. Here we have collected together many of the features to be found singly in other river villages, notably the weather-worn look about the small irregular houses, probably due to the damp atmosphere, and, though not exactly an attraction from the house-hunter's point of view, yet a most desirable feature in the eyes of artists. No crudity can long exist by Thames side; with gentle fingers the soft atmosphere caresses the hard red brick and adds a touch of lichen here and there, and straightway the wall becomes a thing of beauty. Added to this, this same atmosphere, aided by the rich soil, possibly at one time part of the river bed, produces creepers in profusion in every nook and corner; and those asperities which will not yield to gentler methods are veiled by climbing clematis, by masses of wistaria, or by the stretching withy branches of rose bushes. The result is a sweet vista of glory in flower-time, a glory out of which peep casement windows, gable ends, and irregular angles. Roses and sweetbrier, purple clematis and starry jasmine, tall garden plants, and delicate overhanging mauve blooms of wistaria, looking like rare coloured bunches of grapes, mingle with or succeed one another from spring to autumn. The prolific growth in Thames village gardens is one source of beauty to the river. In autumn no strip of a few square yards but has its tall hollyhocks, its royal sunflowers, and, in gay carpets, its scented stocks. The gardens of the lock-keepers, often situated on small islands, are among the gayest on the river; a prize is offered every year for the best of them, a prize which, I believe, Goring has carried off frequently. Matthew Arnold must have had some of these cottage gardens in his mind, when he wrote:

> Soon will the musk carnations break and swell,
> Soon shall we have gold-dusted snapdragon,
> Sweet-william with his homely cottage smell,
> And stocks in fragrant blow;
> Roses, that down the alleys shine afar,
> And open jasmine-muffled lattices.

Besides its flowers and its general architecture, Wargrave has other claims to rank as a typical Thames-side village. The old inn, The George, whose lawn runs down to the water, is just the kind of hostelry one expects to find. Its signboard, indeed, was painted by two R.A.s, a fact eloquent of the kind of "wild-fowl" which forgathers at Wargrave. This unique sign is preserved indoors, while an understudy swings out over the village street.

Wargrave church, too, is no whit behind expectation. It is of flint, as are the most part of the Thames-side churches, and has a square tower with pinnacles, half ivycovered; so it acts up to all that is required of it. Thomas Day, the author of *Sandford and Merton*, which so delighted the last generation of children, is buried in the church; he was killed by a fall from his horse. To add to the list of its self-respecting virtues, the tower of Wargrave church can be seen from the river, peeping out from among the tall trees that surround it.

Above Wargrave is Shiplake, between which and Sonning is the curious channel known as the Loddon and St. Patrick's stream. These two, making a loop by which the lock may be avoided, are tempting to boatmen, for nowhere else on the river may such a feat be performed. Yet if the boatman try the passage up-stream it is likely he will regret it and wish he had favoured the lock, with all its bother and its unwelcome toll instead; for St. Patrick's Stream has a swift current.

Of Sonning who can write with sufficient inspiration? The wonderful old red-brick bridge has drawn artists by the score, whereupon they have drawn it in retaliation! The hotel rose garden, famous for the variety and beauty of the blooms, is an attraction only second, and the hotel itself is second to none on the river.

The mills on the Thames might well have a book to themselves; they are so ancient and so picturesque. Several, including the one at Sonning, are actually mentioned in *Domesday Book*. They are more ancient in their establishment even than the records of the monasteries, and so can claim to be the oldest things on the river, though some of the bridges might run them close. In the hot summer days the backwater of a mill is a place beloved of many. There, beneath the shelter of a broad-leaved horse-chestnut, so thick and rich of growth it makes the water almost black, one may lie in still content, hearing the splash of the falling water, and perhaps seeing it dashing from the mighty flaps of the wheel in glittering cascades. The very sight helps to keep one cool.

Of bridges, too, much might be said, and yet records are hard to find. Sonning bridge must rank high in age, as also that at Abingdon, of which we read:

SONNING

King Herry the Fyft in his fourthe yere,
He hath i-founde for his folke a brige in Berkschire
For cartes with cariage may go and come clere,
That many wynters afore were mareed in the myre.
Culham hythe hath caused many a curse,
I-blessed be our helpers we have a better waye,
Without any peny for cart or for horse.

—*Geoffrey Barbour.*

The building of bridges was in old days considered an act of charity, in the same way as the founding of almshouses and "hospitals". People left bequests with this object.

Between Reading and Wallingford are two other noted beauty bits, which could not be omitted in any book on the Thames, however limited the space. Mapledurham, with its beautiful little church, its fine old Elizabethan house near by, and its most delightful mill, is visited by everyone who can make the pilgrimage. It is, however, rather spoilt by the near neighbourhood of Reading, which is the only town which can be called such, in the real "towny" sense, between London and Oxford. Yet Reading is not exactly on the riverside, but has a river suburb at Caversham. Henley, Wallingford, Abingdon, and the rest are so thoroughly in accordance with the spirit of the river, so charming in themselves, and above all so comparatively limited in extent, they add to rather than detract from the Thames scenery. Reading, in spite of its undoubted features of interest, in spite of its ancient history, is still a manufacturing town, and as such spreads around an atmosphere which is uncongenial to true Thames lovers, who regard it as a blot.

The abbot of Reading was mitred, and ruled with a powerful hand; indeed, the abbey over which he held sway was third in England, and had the privilege of coining, a royal prerogative. Adela, second queen of King Henry I, is buried here, also his daughter the Empress Maude. When the Dissolution came, the abbot in office, Hugh Farringford, thirty-first of his line, nourished on the proud traditions of his predecessors, refused to yield to Henry VIII, and was in consequence hanged, drawn, and quartered in front of his own gate.

There was a castle in Reading as well as an abbey, though the only reminiscence of it left is in the name of Castle Street. From the time of the Danes the castle played its part in history; in the Civil Wars it was at first a stronghold for the King and later for the Parliamentarians. St. Giles's Church still bears the marks of the artillery from which it suffered. Archbishop Laud was born at Reading and educated at the Free School there. At present, as everyone knows, Reading is renowned for its biscuits and seeds.

Farther up we have a repetition of twin villages,

linked by a bridge, veritable Siamese twins, a fact which is interesting and curious. Pangbourne and Whitchurch dwell in the same sort of amicable rivalry as do Streatley and Goring. They may be at war between themselves but they hold together against the world.

Streatley certainly cannot fail to yield the palm to Goring for beauty. For Goring is considered by many critics to be the very prettiest village on the river, a claim which its quaint main street, falling down the hillside to the river at right angles, does much to establish. But the surroundings of Streatley, the splendid sweep of heights, which back it up, cannot be rivalled by Goring. The road running through both crosses the river, and it is ancient in very truth. It was used by the Romans and formed part of the famous Icknield Way, but was made long before their time. For generations before history begins bands of furtive men, ready for surprise, and as suspicious as wild animals, must have padded on bare feet down one line of hills, across the river ford, and mounted the heights again, keenly scanning the country for possible enemies. No neat creeper-covered red brick cottages then, no church even, though Goring church is very old, dating back to Norman times, and having been the church of an Augustinian priory. No mills even, not the most primitive, and though neither village can be accused of ruining its beauty in a frantic search after modernism—the mill at Goring, in spite of its mossy roof, gleaming green and russet, frequented by the flocks of white pigeons, has adopted an electric generating station! From the electric-power methods to the Ancient Britons is indeed a far cry!

Pangbourne and Whitchurch, taken as a couple, cannot vie with Goring and Streatley; though Pangbourne is pretty enough, and the river near it is island-broken, and particularly attractive. The reach succeeding Goring and Streatley is dull right up to Wallingford. In some points Wallingford and Abingdon may claim brotherhood, they are of the same size and about them hangs the same atmosphere, but the river at Abingdon is incomparably more interesting. Of Wallingford something more must be said in the historical reminiscences, and for the time we may leave it, and, skipping Dorchester, already mentioned, and Sutton-Courtney, another beauty spot, with an incomparable "pool", go on to Abingdon.

Of the bridge we have already spoken—there it stands, Burford Bridge, old and irregular, with straggling arches, some round, some pointed. The bridge is long and rests partly on an island on which is built the Nag's Head Inn, whose garden occupies the island. The abbey buildings, still partly standing, founded by Cissa in 675, is one of the most interesting features of the town. The long range of wall, and the mighty exterior chimney, probably built about the fourteenth century, show up in season amid masses of horse-chestnut blossom, for which the town is famous. Henry I, the learned Beauclerc, was here educated from his twelfth year.

Christ's Hospital, as it is called, with a hall dating from 1400, is one of the sights of Abingdon, and the day to see it is that on which eighty loaves of bread are distributed to the poor people of the town. This occurs once a week.

With Abingdon we get within range of Oxford, and what remains is distinctly in the Oxford zone, just as all the river below Hampton is London in character. The famous Oxford meadows, with their range of wild flowers, rival the Swiss meadows.

The profusion of flowers in the riverside gardens has already been noted, but these differ little, except in richness of growth, from those usually found in cottage gardens. More interesting to those studying the Thames as a theme are the flowers growing wild along the banks, which are native to the river. Among these may be reckoned the purple loosestrife, with its tapering gaily coloured spikes standing often four feet high, and at times mistaken for a foxglove; also the pink-flowering willow-herb, the wild mustard with its raw tone of yellow, the buckbean growing in low-lying stagnant places, and the tall yellow iris, clear-cut and soldierly, with its broad-bladed leaves rustling along the margin of the banks. Not less beautiful are the burr-reeds and flowering rushes, the marsh-mallows and the cuckoo-flowers, found in many parts of the river; but the growth of wild flowers, including these and others, is richest of all in the meadows below Oxford. Here the fritillaries are especially noted:—

> I know what white, what purple fritillaries,
> The grassy harvest of the river fields
> Above by Ensham, down by Sandford yields.
>
> —*Matthew Arnold.*

Also the yellow iris, the cuckoo flower, the water villarsia, the purple orchis, the willow-weed, and many another are here seen in full perfection. The Nuneham woods rank with the Oxford meadows as an attraction, and the inn at Sandford still holds its own, though overshadowed by a paper mill.

There is one glorious gem by the river which is in a category by itself, and is unapproached by rivals; this is the small church of Iffley. Its architecture is not pure, but its claim to date from Norman times is undisputed. No one passing along the meadows should fail to stop at Iffley and see some genuine Norman mouldings and massive architecture.

After this we come to Oxford and may stand on Folly Bridge, and as we watch the water flowing swiftly beneath our feet may run with it in imagination past all the beauties and all the places of interest already described, on by cool meadows and overshadowing trees until it meets the flooding uptide below Richmond and mingling with it in the ebb is lost in the "town" water of Brentford and Hammersmith, and so plunges into the thick grey flood by London, and on by wharves and docks until—

> Stately prows are rising and bowing,
> Shouts of mariners winnow the air,
> And level banks for sands endowing
> The tiny green ribbon that showed so fair.
>
> —*Jean Ingelow.*

No river in the world can show so wonderful a gallery of great names, or so noted a collection of world's men, in connection with it. Perhaps the two names which arise at once to everyone's mind are those of Pope and Walpole, who lived so near one another at Twickenham. Pope was at Twickenham from 1719–44, and produced here his most famous works, including the last books of the *Odyssey*, the *Dunciad*, and the *Essay on Man*, but he is not by these remembered on the river, his claim to notice is that he made a curious underground grotto, of which he wrote:—

> From the River Thames you see through my arch up a walk of the wilderness to a kind of open temple, wholly composed of shells in the rustic manner, and from that distance under the temple, you look down through a sloping arcade of trees, and see the sails on the river, passing suddenly and vanishing as through a perspective glass. When you shut the doors of this grotto it becomes on the instant from a luminous room a camera obscura, on the walls of which all objects of the river, hills, woods, and boats, are forming a moving picture in their visible radiation.

Pope had known the river from his birth. His parents lived at Binfield, about nine miles from Windsor. Part of Windsor Forest is still called Pope's Wood, and his poem on Windsor Forest must contain some of his earliest impressions. He was two years at Chiswick, after leaving Binfield, and then bought the house at Twickenham with which his name is chiefly associated. Long before this, however, he had been a popular visitor at Mapledurham, where the glorious old Elizabethan mansion near the church still shelters Blounts as it did in his day and long before. Two pretty daughters of the house, described by Gay as—

> The fair-hair'd Martha and Teresa brown.

competed for the honour of Pope's attentions, even though he was "a little miserable object, so weak

PANGBOURNE

that he could not hold himself upright without stays, so sickly that his whole life was a continued illness"; his genius, early recognized, concealed by its blaze such trifles. His poems in many places keep alive the sisters' names, and in the Mapledurham MS Collection much of his correspondence is preserved. There does not seem to have been any question of his marriage with either of the girls, and it is doubtful if his connection with them was altogether for their good; but at any rate it has added lustre to the family records. Teresa once assured him, he tells us, "that but for some whims of that kind (propriety) she would go a-raking with me in man's clothes".

One detail of Pope's garden is so peculiarly associated with the river that it must be mentioned. It is said that the weeping willow grown by him was the parent of all the weeping willows in England, and if so many a Thames vista owes an added touch of beauty to him.

Pope's grotto has taken so much hold on the popular imagination that it ranks only second to his hideous and grotesque villa by the riverside, which was recently occupied by Henry Labouchere, M.P. The real interest of the place lies in the literary coteries which met in the house, including such men as Swift and Gay, who helped by suggestions and designs during the building of the famous Marble Hill for the Countess of Suffolk, friend of George II. Gay in particular was a *persona grata* with the countess, and occupied a special suite of rooms set aside for him at Marble Hill.

It was three years after Pope's death that Walpole came to the neighbourhood; he had the mania for fantastic building effects even more strongly than the poet. Pope had made his villa peculiar enough in all conscience, but Walpole's so-called Gothic in the rebuilding of Strawberry Hill was a medley of every sort of architectural effect which could conceivably be classed under that heading. "Not to mention minute discordances, there are several parts of Strawberry Hill which belong to the religious, and others to the castellated, form of Gothic architecture." Walpole solemnly boasted that his "house will give a lesson in taste to all who visit it". It might have done so, but not exactly in the way he intended. He made the place a perfect museum, and it became the fashion to visit Strawberry Hill. The Earl of Bath was so enchanted with it that he wrote a ballad, which, in its own kind, might well take rank with the architectural effort which inspired it. Every verse ended:

> But Strawberry Hill, but Strawberry Hill
> Must bear away the palm.

Walpole wrote of the place, soon after he had acquired it: "Two delightful roads, which you would call dusty, supply me continually with coaches and chaises, barges as solemn as barons of the exchequer move under my window. Richmond Hill and Ham walks round my prospect; but, thank God! the Thames is between me and the Duchess of Queensberry!"

He used to term the mansion his "paper house" because, the walls being very slight, and the roof not very secure, in the heavy rains it was apt to leak, "but," adds an enthusiastic writer of his own time, "in viewing the apartments, particularly the magnificent gallery, all such ideas vanished in admiration".

After his first visit to Paris, Walpole never wore a hat, and used to go out walking over his soaking lawns in thin slippers. He sat much in the breakfast-room, which gave a view toward the Thames, and his constant companion was an inordinately fat little dog. He wrote the *Castle of Otranto* in eight days, or rather eight nights, for he says his "general hours of composition are from ten o'clock at night till two in the morning".

The squirrels at Strawberry Hill were a great feature; regularly after breakfast Walpole used to mix a large basin of bread-and-milk and throw it out to them. He was very fond of animals, he even used to cut up bread and spread it on the dining-room mantelpiece, thus drawing a number of expectant mice from their holes!

It troubled him greatly when he became Earl of Orford, at the advanced age of seventy-four, on the death of his nephew. He could not see why, sitting at home in his own room, he should be called by a new name!

The most notable fact connected with Strawberry Hill was the printing-press Walpole there established, from which he issued many of his own, and some of his friend, the poet Gray's, works.

Henry Fielding came to Twickenham, having first married, as his second choice, his late wife's maid. He was only here about a year. Sir Godfrey Kneller, too, was a resident; and Turner, having built here a summer resort, and called it Sandycombe Lodge, used it from 1814–26. So that, all things considered, Twickenham may boast a considerable galaxy of stars.

Though the names of Pope and Walpole are best known from their long association with the river, by far the noblest name that Thames can boast is that of Milton. It was as a young man, fresh from the University, that he came to live for five years with his parents at Horton, near Wraysbury. Horton is not exactly on the river, but it is very near, and

FOLLY BRIDGE, OXFORD

the influence of the scenery must have been strong on the delicate youth nicknamed "the lady", whose genius was already blossoming. He walked far and wide over the rich, well-watered land, down to the river's banks with its overhanging trees. In many of his stately poems little word pictures, reminiscences of these quiet days, are found:

By the rushy-fringed bank
Where grows the willow and the osier dank.
—*Comus.*

Ye valleys low, where the mild whispers use
Of shades and wanton winds and gushing brooks.
—*Lycidas.*

The house in which Milton lived has vanished, in fact the only one of his many residences remaining is that at Chalfont St. Giles, Bucks. But the pretty little church at Horton, close by which the house was situated, still stands. The poet's only sister was married, his younger brother an occasional visitor, and, as his father was well on in years, the life must have been singularly quiet. Milton was only in his twenty-fourth year when he left the University, but already his poems had shown the bent of his mind. He was at Horton from 1632–38, and he himself says he spent there "a complete holiday in turning over the Greek and Latin writers". Hardly the kind of holiday that would commend itself to the Etonians not so many miles off. Yet this "holiday" was productive of *L'Allegro*, *Il Penseroso*, *Arcades*, and *Comus*, all ranking among the greatest classics in the English language.

It is in single lines the effect of the landscape he knew best is seen.

> By hedgerow elms on hillocks green.
> Meadows trim with daisies pied,

are redolent of the Thames country. Milton's mother died in 1637, and was buried in Horton Church: soon after the poet went abroad.

Another poet of the first rank who may be claimed by the Thames is Shelley, who was at Great Marlow when he wrote *The Revolt of Islam* and *Alastor*. The cottage is now divided into four and is easy to see, as there is a long inscription, giving details about the poet's occupation, upon the front of it. *The Revolt of Islam* was written partly as he sat in the Quarry Woods and partly in a boat; so it belongs peculiarly to the river.

Matthew Arnold has already been mentioned, and many of his poems show strong impressions of the river scenery. He was born and is buried at Laleham, where his father, the afterwards famous Dr. Arnold of Rugby, had settled down to take pupils for the Universities.

STREATLEY HILLS

Another name the Thames can claim is that of Cowley. The house in which he lived for two years before his death in 1665 is still standing, at Chertsey.

It is easy to see, therefore, that the river can boast more poets of high rank than any other celebrated men. This makes it the more peculiar that there is no great poem on the subject.

Above Molesey Lock, at Hampton, stands the house bought by the great actor Garrick in 1754. The place is known better by the little Shakespeare Temple near the water than by the galaxy of great names drawn thither by Garrick himself. We have in Fitzgerald's *Life of Garrick* a living picture of the daily comings and goings; we see Mrs. Garrick discussing laurel cuttings with the Vicar, or eating figs in the garden with her husband, who was dressed in dark-blue coat with gold-bound buttonholes. At all sorts of odd hours Dr. Johnson burst into the family circle, and when consulted as to how best the ridiculous little "Temple" could be reached from the house, from which it was divided by a road, broke out in all earnestness in favour of a tunnel, as against a bridge, in the words: "David, David, what can't be over-done may be under-done!" One terrible night, when the sensitive actor read aloud from Shakespeare, his guest, Lord

March, fell asleep. The sting was the deeper as "Davie" dearly loved a lord! The river fêtes Garrick gave were renowned, and the fame of them remains to this day; alas, the knack of river pageantry has long been lost!

Carlyle, in later days a frequent visitor to the villa, once drove a golf ball through the centre of a leafy archway clean into the river.

History is notoriously dull, except to those who have a taste for it, but yet there are scenes in history which may stand out as brightly as any pictures. Of such is the signing of Magna Charta, the greatest act recorded in the whole of our English annals. Well might it be thought that London, by means of the Tower or Westminster, would have claimed to be the theatre of so epoch-making a scene; not at all; as the youngest child knows, it was no building which witnessed the deed, but a Thames-side meadow, which may be seen to-day all unchanged, and happily as yet unbuilt on. The island, which goes by the name of Magna Charta Island, is now generally supposed to have usurped a claim properly belonging to the meadow by Thames side, and we confess to a certain pleasure that this discovery has been made; for the island is altogether too trim, too neat, and the house thereon too modern, to assort with thoughts of a mighty past. No, we who love the river believe rather, and in our belief we are backed by the latest research, that the flat land, encircled by the heights of Cooper's Hill, as by the rising tiers of seats, was the amphitheatre whereon the great scene was enacted. We can imagine it crowded by mailed men who trampled under foot the mushy grass, mushy even in the season of summer, an English June. The exact date, never to be forgotten, is June 15, 1215.

The flowers grow well about here, the spotted knotweed, the common forget-me-not, the pink willow-herb, the yellow iris, and purple loosestrife may all be found in season, and the meadowsweet and dog-rose scent the summer air.

Everyone knows about Magna Charta, but few perhaps realize that Kingston has an older historical claim than Runnymeade, for it owes its name to being the seat of government of our oldest kings. In the marketplace may be seen the stone inscribed with the names of the seven Saxon kings here crowned in turn; hence Kings' Stone. At that date Mercia and Wessex were united under one king, and the boundaries of Mercia came down to the Thames on the north side, while those of Wessex marched with them on the south. London was unsafe because of the ravages of the Danes, and as at Kingston from time immemorial there has been a ford, a thing of vast importance in the absence of bridges, and a ford well known, it seemed that Kingston had some claim to the ceremony. In 1224 a wooden bridge replaced the ford, the oldest bridge, and the only one, between this and London Bridge. The bridge itself has played a historic part. In 1554 Sir Thomas Wyatt, marching to London, found London Bridge closed against him, so he had to march as far as Kingston to reach the next crossing-place. The fact seems incredible to us in the days of many bridges. But when Sir Thomas arrived at the end of his tedious march he found he had been forestalled, the bridge was broken down, and on the farther bank two hundred soldiers stood ready for him should he dare to use the ford! Therefore back went he to London Town.

Wallingford has a little bit of history of its own. It boasts the oldest corporation in England, a hundred years prior to that of London. It also disputes with Kingston the claim to the oldest bridge and ford above Westminster. The town was "destroyed" by the Danes in 1006. At the time of William the Conqueror's advance on London the castle was held by Wigod, a Saxon, and from that time onward it was a notable fort, taking part in many historical events. It boasted three moats, and a fragment of the old wall remains in the pretty garden of the house now

WALLINGFORD

called the Castle. In 1153 Prince Henry "lay" at Wallingford with 3000 men, and Stephen, with another army, glared at him from the opposite bank; but like two schoolboys, mutually unwilling, the rivals slipped away without encounter. It was Cromwell who ordered the utter destruction of the castle in 1652.

The oldest historical incident of all in connection with the Thames is the supposed crossing of Cæsar at Cowey Stakes, above Walton Bridge. Some strong wooden stakes, black and tough with age, and metal-capped, were found driven into the bed of the river at this point. They are supposed to have been driven in by the Britons to hinder the crossing of Cæsar in B.C. 54. As it is known that Cæsar did cross the river some eighty miles above the sea, and as a Roman camp was discovered in the neighbourhood, it is quite possible that anyone standing on Walton Bridge, looking over the wide peaceful stretch of river above, is really surveying the stage on which one of the earliest acts in our great national drama was played.

The unhappy Henry VI, too weak to bear without misery to himself the responsibility life thrust upon him, sleeps at Chertsey. His body, after being exposed at Blackfriars, was brought here on a barge—a slow procession and a sad one. In *Richard III*

Shakespeare makes the hyprocritical Duke of Gloucester say:

> After I have solemnly interred
> At Chertsey monastery this noble king,
> And wet his grave with my repentant tears.

Not far from the resting-place of Henry VI, a great statesman, Charles James Fox, was born. What a gap in time and manners and customs is here suggested. To think of the two is to span the distance between generations of growth and thought. Fox died at Chiswick House, so his life began and ended by Thames side. In the same house, twenty years later, died another great statesman, George Canning. Thus, even without reckoning London itself, the centre of our national life and history, we find the Thames can show names famous in literature, in history, and in politics. Its banks are studded with memories as they are with flowers, and in contemplation and reminiscence the annals of the centuries flow past us as the water itself flows by, ever smoothly and unceasingly.